Chromecast User's Manual Streaming Media Setup Guide with Extra Tips & Tricks!

By Shelby Johnson

Disclaimer:

This eBook is an unofficial guide and companion manual for the Google Chromecast Dongle, and it should not be considered a replacement for any instructions, documentation or other information provided with the Chromecast device or by Google on their associated websites. The information in this guide is meant as recommendations and suggestions, but the author bears no responsibility for any issues arising from improper use of the devices mentioned within. The owner of the device is responsible for taking all necessary precautions and measures with their TV system and Chromecast Dongle.

Images and screenshots were acquired from Google Chromecast setup pages, Chrome browser, Amazon, or taken by the author, and they imply no affiliation with Google or the Chromecast device or Amazon.

Who This Guide is For

Perhaps you're an owner of the new Chromecast dongle by Google. Or maybe, you are considering purchasing one. While this guide is not affiliated or associated with Google or the creators of Chromecast, it can help you whether you're an owner, or a prospective buyer, who wants to learn more about the technology and how it can be used!

Inside this guide, you'll find overviews of how to set up the Chromecast with various devices. You'll also find a helpful list of other websites you can use to stream content from to your TV or display monitor. In addition, you'll find ways to tweak your home network setup, or your streaming computer in order to maximize the speed and overall playback quality of streamed content. If you're a prospective buyer, this guide may very well help you understand what you are getting, and what else you might need to have in order to fully utilize this new technology.

Contents

Introduction

Perhaps you're an owner of the new Chromecast dongle by Google. Or maybe, you are considering purchasing one. While this guide is not affiliated or associated with Google or the creators of Chromecast, it can help you whether you're an owner, or a prospective buyer, who wants to learn more about the technology and how it can be used!

Inside this guide, you'll find overviews of how to set up the Chromecast with various devices. You'll also find a helpful list of other websites you can use to stream content from to your TV or display monitor. In addition, you'll find ways to tweak your home network setup, or your streaming computer in order to maximize the speed and overall playback quality of streamed content. If you're a prospective buyer, this guide may very well help you understand what you are getting, and what else you might need to have in order to fully utilize this new technology.

Sign up today for our mailing list and receive updates as Google adds more features to its Chromecast device.

What's In the Box?

Chromecast is a very simple media streaming tool, which is evident from the contents of the box it comes in. There are four items in the box, and even the box itself is a sleek, sturdy and functional tool that allow for crush-free storage of the components you will not be using.

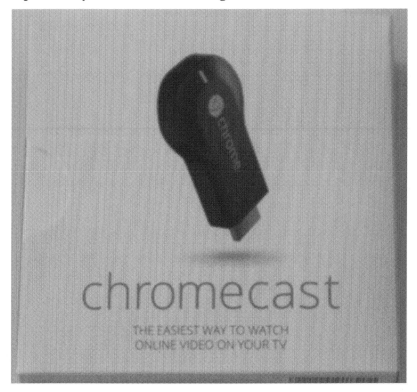

Google Chromecast

Chromecast is weighs just over one ounce (1.1, to be exact), and measures a mere 2.8 x 1.4 x 0.5 inches. It looks like a data or jump drive, and is officially referred to as a dongle – which is basically a device that connected to a computer to allow access to wireless broadband. It is small like a removable drive, and plugs in like one as well. You simply plug it into your HDMI port on your television, and it connects to your existing Wi-Fi for use with iOS devices (iPads, iPhones), Android devices (smartphones and tablets), and even your laptop. It must connect to a Wi-Fi enabled device, which means that if you are still using an old desktop computer, this device is not going to do you much good.

It allows you to stream media directly from any of these devices wirelessly, including movies, television shows, pictures, and music much like Apple TV and Roku do. The difference is – among other things – it is really small, and does not require the use of a remote.

USB Power Cable

The USB Power Cable is what powers the Chromecast device. One end plugs into the device, and the other end plugs into the USB port on your television, or your power supply (if you do not have a USB port on your television – see below). This is absolutely necessary, as the device is not going to power itself. Some have been turned off by it needing a power source at all, which is probably the result of its small, drive like appearance. When you take it out of the box, it appears to be self-sufficient, so having to plug it into something else gives techies a bit of a scowl. In reality, it simply is not that big of a deal. Everything needs power.

Power Adapter (Optional Use)

If you do not have a USB port on your television, the first thing you will figure out is that you have not upgraded in a while, huh? Flat screens are getting less and less expensive each year, so go ahead and splurge! Uh hum. Sorry. If you do not have a USB port on your television, you will need to plug the USB Power Cable into your Chromecast on one end, and the power adapter on the other. The power adapter, as the name would suggest, will plug directly into the wall.

HDMI Extender (Optional Use)

The HDMI Extender is also optional, and helps you connect the Chromecast even if does not plug directly into your television. It is, by all accounts, an extension. That is to say, if you cannot reach the back of your television (if it is mounted to the wall, perhaps), the extension will give you the room you need to still use the device. It may also improve your Wi-Fi reception, so hang onto it in case your television prompts you to use it during set up. If you do need it, connect the HDMI extender into the television first, and then plug the Chromecast into the extender. This will keep you from damaging the small device before you even get to use it!

Small Instruction Manual

There is a small instruction manual that comes in the box as well. Small describes this instructional sheet perfectly because it appears to have been designed and printed for a baby doll to read. It is so tiny, in fact, that it is hard to turn the pages without feeling silly. The good news is, the Chromecast is simplicity at its best, so the manual isn't really necessary. But it is cute.

What You Will Need to Use Chromecast

You should already have some sort of television or display monitor which has an available HDMI port to plug the Chromecast into. If it has a USB port, that's also helpful for plugging in the micro-USB power cable to. If your TV doesn't have the USB port, don't worry, you can plug the micro-USB cable (included) into the included USB/AC power adapter, and then plug that into a nearby wall outlet.

You'll want to have a wireless network set up in your home (with a wireless modem and possibly a wireless router). If you have a secure wireless network, make sure you have any password or other info handy to enter for the Chromecast setup.

For streaming content to your Chromecast on your TV or monitor, you'll need a Mac, Windows or Chromebook Pixel computer. Additionally, you can use an Android or iOS smartphone or tablet. At this time, only certain apps will work on the mobile devices for streaming content. These include YouTube and Netflix, which you'll be able to use. In the future, there is very likely to be more apps developed that will allow you to use the Chromecast streaming to your TV.

Setting up the Chromecast

Google makes it pretty clear that they would only like for you to use the accompanying USB cord to operate the device. As nearly everyone has extra USB cords lying around, Google makes it a point to say, "Use Ours" in their directions, and in any online advisement. It certainly cannot hurt to head their advice, and besides – it comes with the device, so you may as well use it for the purpose it was intended.

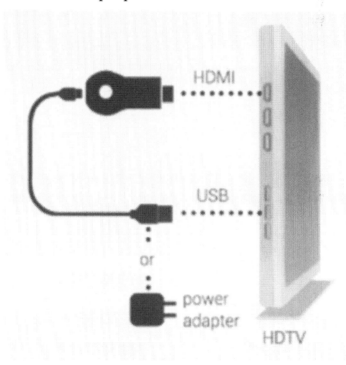

Setting up the Chromecast is really quite simple. In fact, more so than most technological devices, which really is a source of pride with Google, and completely reflective of their brand. From the moment you take it out of the box you will immediately be wowed by its composition and user friendliness. The set up takes a few steps, and will require a smidgen of patience, but it is completely painless. Do not be fooled by its tiny stature, however. This dongle provides exceptional usability, and you will love its addition to your technological cache as soon as it is set up.

Connecting to the TV

This device is so easy to use that all you have to do is plug it directly into the HDMI input on your television set. If you need to use the extension for a better fit or to increase your Wi-Fi reception, plug it in first, and then plug the Chromecast into the extension. That is it. You will see a message on your television screen that says, "Set Me Up." But before you can go through the set up process, you have to connect it to your network Wi-Fi first.

Connecting to Wi-Fi

Since the Chromecast does not come with a remote, like the Apple TV and Roku does, you are going to have to begin the set up process and Wi-Fi connection using your computer or tablet. In fact, your computer, tablet or smartphone – or a combination of any of those – will serve as your remote at all times, which is a refreshing alternative to learning something new, or to losing the remote altogether.

Although it is not impossible, your electronic devices typically stay within range, and are not as easy to misplace as, say, a small media streaming device remote. You already know how to use your existing devices, so combining the two technologies is effortless, and a welcomed change from adding yet one more remote to your already non-functioning versions in a basket atop your entertainment stand.

In order to connect to Wi-Fi you will logon to the Google Chromecast set up page at www.google.com/chromecast/setup, and start going through the steps provided on your computer, tablet or smartphone's screen. It will ask you if you see the "Set Me Up" message on the screen, and you will click "Yep, got it" assuming you do "got it", that is.

Next it will ask you to switch your Wi-Fi network so it matches your Chromecast name, and once it detects the device it will display a four letter/number combination code on the television screen, like C4S2 (not exactly that, but a combination that is similar in character) to confirm that the device you are using is the correct one for the network. This is to avoid any confusion where multiple Wi-Fi options are available, like in an apartment building or office. You will then switch back to your original Wi-Fi connection and begin using your new Chromecast to your heart's desire in no time.

You will be prompted to enter your Wi-Fi password for verification when you switch back to your regular network, and afterwards your television screen with say, "Ready to Cast." Yay! You are all set to go, and begin streaming from your electronic device directly to your television set. This is all very exciting and to be honest the Google Chromecast, upon set up, gives you a tremendous amount of power to stream anything and everything you could imagine from your existing device, so get ready to play with a lot of content in your first sitting.

Using the Chromecast with PC

First things first: You are going to need Chrome for Windows on your PC in order to use the Chromecast device effectively. In fact, once you have the internet browser, you are also going to want to install the Chromecast extension as well. You will be prompted by the forward thinking Google programmers to do so, and you will be glad you did.

You can also use the supported "apps" which are Google Play, YouTube and Netflix, which are technically downloads on your computer more than they are apps per se, to view content seamlessly from your PC to your television. You can also use your actual browser tabs as content, which means Pandora and Vimeo are easily transferred onto the larger screen too. So is your favorite shopping outlet or blog. So, in effect, you are mimicking your computer's browser onto your television screen. This is a great way to read small type, and to share the PC's browsing capabilities with the rest of the room. So if you want to go out for a movie later, you can pull up each of the trailers in your browser, and watch them on your television before making a decision. It's genius, really.

Using the Chromecast with Mac

Much like the PC's use of Google Chromecast is a necessity, the Mac operates almost exactly the same way. You will need Chrome on your Mac, and it will prompt you to download the Chromecast extension as well. You can stream from the available applications like Netflix and YouTube or simply use your browser to stream the material just as you would while watching it on your laptop.

The big drawback seems to be – so far anyway, which is to say this is the first iteration of this streaming device – unlike Apple TV or Roku, you will not see the apps on screen, nor will you be able to choose from a selection. You will have to do everything from your Mac by physically going to the page in question through your Google Chrome browser, and streaming the media from there. What will appear on screen until then is a landscape image that provides no direction whatsoever, so you almost feel as if you did something wrong. You did not. You just need to control the content from your computer instead of expecting Chromecast to do it for you.

Using Chromecast With a Mac Mini

The Mac Mini computers by Apple are compact and small but pack plenty of punch. If you have one of the newer ones in your home somewhere and it is part of the wireless network, you can use it to stream to a Chromecast device (depending on how good the wireless is in your home). I personally tested mine from the basement of my home and it was able to stream to a TV upstairs in my living room.

To set up Chromecast on the Mac Mini complete the following:

1. Go to the google.com/chromecast/setup webpage.
2. Download an app (.dmg file) to your Mac Mini as directed.
3. Double click the app when it is done downloading. A pop-up box will appear directing you to drag the Chromecast app to your Applications folder.
4. After you drag the app to that folder, double click the Applications folder to open it, and find the "Chromecast" app in the listed files. Double click on the file to open it.

The Chromecast app will open. Make sure you have your TV on with the Chromecast powered, so you can connect your Mac Mini. You'll be able to connect to your Chromecast and adjust settings within the app.

You'll also be directed from there to add the Chromecast browser extension which will help you stream from Google Chrome browser tabs on your Mac Mini.

Once done, you can close the app and start using Chromecast with your Google Chrome browser on the Mac Mini!

Using the Chromecast with Tablets (or Smartphones!)

Okay, now you are talking. Although it is perfectly fine and dandy to run the Chromecast from your PC or Mac, having a tablet or smartphone – whether it is iOS (Apple) or Android technology really allows your streaming capabilities to come to life.

First, tablets and smartphones operate with apps, which is a beautifully easy way to access the device's content. Netflix and YouTube apps are easily mirrored on your television screen by simply tapping their icons (one at a time of course, this device isn't magic) on your tablet or smartphone (you will have to download these apps if you do not already have them...it still isn't magic). So if you are a Netflix member, all you have to do is tap on the app's icon, and start flipping through your favorite shows and movies just as you would if you were watching them on the tablet. See? Nothing new to learn! Just do what you have always done, except on a much larger screen. That means you and your friends will not have to huddle around your iPad to watch last night's episode of whatever television series you are into.

When using YouTube apps on any supported tablet or smartphone, you'll notice the small cast icon on the video (usually up near top right area of video). Click on the cast button and it will bring up the option to cast to your Chromecast, displaying the YouTube video up on your TV screen.

To end the casting session from your device, simply click on the cast button again and choose to show the video on your device.

Supported Technology

Here is what you will need, no matter which device you are running the Chromecast from: Chromecast is compatible with Wi-Fi enabled Android 2.3+ smartphones and tablets; iOS 6.0+ iPhones, iPads, and iPods; Chrome for Mac and Chrome for Windows; and Chromebook Pixel.

Collaborative Play

There is something inherently exciting about collaborative play that accelerates the senses. Think of the people you have over to your house, and what types of devices they have. With Apple TV you can only stream onto your television from your Apple devices through AirPlay. So if you wanted to share your pictures on the big screen, you better have matching the technology. That does not apply with Chromecast.

Much like anyone with a smartphone or tablet can stream their music to a wireless speaker simply by connecting, anyone can stream their viewing goods onto your television too. This is perfect if you know someone with a Netflix account, as you can watch movies on their dime. It is simple to connect, and instead of using AirPlay, you "cast" the video onto the screen using a similar button that will appear on YouTube and Netflix when the device is enabled.

Pause, Play & Rewind

Just as you are able to do while watching movies or television shows on your tablet or phone, you can hit pause, play and rewind whenever you want when you are streaming onto your television through Chromecast. Since your electronic device becomes the remote, all of the technology transfers instantly, so you are not forced to run to the kitchen and miss pieces of your programming. Simply pause and resume as you wish!

Chromecast Updates

After your initial setup and first use of Chromecast, you will likely unplug the device from its power source, or shut off your TV/monitor. The next time you start up Chromecast you may notice it performing an update. It is likely from time to time that these updates will take place and bring new features to the device. Allow the update to run for several minutes, and then the device will automatically power off and back on. Then it will be ready to use.

Using Multiple Chromecasts Together

A single Chromecast will only allow streaming to the TV it is currently connected to, unless you move it to another TV. So, for each TV set in your home you want to stream to, you will likely want to have a separate Chromecast dongle to connect to it. Otherwise, if you've only got one Chromecast, you will have to move it from TV to TV when you want to stream to a different monitor. Constantly moving the Chromecast can become a bit tiresome, despite this device's portability.

Multiple Chromecasts will work fine together, allowing individuals to stream programs to different TV's in the home, office or other environment. You will set up each new Chromecast based on the initial setup instructions, connecting to the wireless network, and so forth.

Once you have multiple Chromecasts installed, you can simply choose the Chromecast you want to cast to from your device. Make sure to give each Chromecast a unique and helpful name to find it, such as "Living Room TV," "Master Bedroom," etc.

Keep in mind, as mentioned in the increasing internet speed section, that you will be straining the overall load on your wireless network when you are streaming to more Chromecasts in the home, so the quality may suffer somewhat when you stream to multiple devices at the same time.

Features and Apps

Although Google Chromecast does not come with as many "channels" as Roku, or the number of apps as Apple TV it does have a couple of standbys that will help you enjoy television a little more than you did before you purchased the device.

Because it only costs $35, it really is hard to beat the Chromecast. The good news is, as its popularity soars, so will the app availability.

Netflix

Just as you would watch your favorite television shows and movies on Netflix on your laptop, tablet or smartphone, now you can stream the programming directly onto your television set when you have Google Chromecast.

Just as it is with all Netflix programming, you must be a subscriber to enjoy its content. And for $7.99 a month, you can be. Your Netflix subscription will give you complete on demand access to thousands of movies and television shows for your enjoyment. Thanks to Google Chromecast, you can now stream those movies and television shows directly to your larger screen, instead of counting on your laptop or tablet's smaller display. This is great for group viewing, or movie night with your sweetie. Finally, no one has to be responsible for holding the tablet steady, or burning their thighs with a laptop sitting on top of you for the duration of the programming.

YouTube

If it no secret that you can spend hours searching videos on YouTube, with only a few keywords like "cute kittens" or "concerts." It is a wealth of information, and even provides access to exclusive YouTube channels where you can watch your favorite independent programming. You can also watch movies, music, sports and gaming while compiling your own online library by signing in using your YouTube username and password, and adding the items to your account. This is a great way to compile a library of films, music and favorite videos without having to search for them over and over, when you are ready to share your findings with others.

You can also shoot video and upload it yourself, if you want the world to see how your garage band has evolved over the years. That is the heart of YouTube's existence, by the way. No one should lose sight of that magic. Speaking of evolving, YouTube is no longer just home to adorable puppies at play videos, as it now offers full length movies just as you would see them from the studio release. It also offers full movie trailers and music videos, as well as product reviews and instructional guides.

YouTube viewers can listen to music, watch concerts, and enjoy international programming all with a few taps of their fingers. And since there is nothing new to watching the content on your tablet or phone, transferring it to the screen is just as easy.

So the next time you want to put that IKEA desk together, and plan to watch someone else do it on YouTube first, stream it on your television and follow along. That is what technology is for, isn't it?

TV Queue

TV Queue is a Chromecast specific feature for YouTube that allows you to make a line-up of YouTube videos that will automatically play on your TV screen. This has been equated to creating your own customized TV channel, and the possibilities for using this fun feature are nearly endless.

Google Play

Google Play is not a surprising offering, considering it is part of the brand that is bringing you the Chromecast. Google Play is the former Android store, which was merged with Google Music in March 2012, leading to the renaming of both as Google Play for branding and distribution purposes.

Google Play, besides being the official app store of all Android devices, allows its users to play music, movies, television shows and games from their app. It also allows books and magazines to be enjoyed at any time, for your reading or listening enjoyment. Instead of enjoying these offerings on your small screen, you can now stream them directly onto the television with Google Chromecast.

It is also a lot easier to find what you are looking for when you have the luxury of using the same device you use for everything else, so you are not scrolling through letters one at a time with a remote, or spelling out words is the most inconceivable way possible. Use your device's keyboard, and get more done.

HBO GO

With HBO GO, HBO subscribers can log in using credentials from an approved cable or satellite provider and watch their favorite HBO original series as well as favorite movies any time without having to worry about the network's programming schedule.

HBO Go includes every episode of every season of the best HBO shows, movies, comedy, sports, and documentaries.

Hulu Plus

With your Hulu Plus subscription, instantly watch hit TV shows using your Chromecast device. Hulu Plus currently offers a free one-week trial subscription, and after that, if you like the service, you can keep it for $7.99 a month. The monthly subscription includes limited advertising, and you can cancel your Hulu Plus subscription any time.

Hulu Plus includes both current season and back seasons of many popular TV shows. Plus, one subscription will not only play using your Google Chromecast device, but it will also play across your other Hulu Plus enabled devices.

Pandora

Chromecast makes it easy for you to listen to your favorite free, personalized music using the best speakers you own. Pandora is free, personalized radio that plays music you'll love. Discover new music and enjoy old favorites. With Pandora you can explore this vast selection of music.

Just enter the name of one of your favorite songs, artists or genres into Pandora, and the Music Genome Project will quickly scan all its analyzed music, which includes nearly a century of popular recordings, to find songs that are similar to your original choice. Pandora will create a station filled with your current and plenty of new favorite songs just for you.

Using the Pandora app with Chromecast, you can create up to 100 different "stations" for your listening pleasure. If something about one of them is not quite right, you can indicate that, and Pandora will improve your station to better suit your musical tastes.

VEVO

Instantly watch your favorite music videos using the VEVO app on your Chromecast. VEVO includes a library of 75,000 HD music videos, exclusive original programming, and live concert performances, all of which can stream straight to your TV using the Chromecast.

RedBull.TV

Experience the inspirational programming from Red Bull. Options include sports, music, and lifestyle entertainment. The collection is extensive, so there is sure to be something to suit nearly any taste.

RedBull.TV offers a fresh escape to destinations throughout the world. This free service allows you to drop into a club in the Caribbean, a party in Brazil, or a surfing expedition in the South Pacific all without leaving the comfort of your own home. Tune in today to see what you have been missing!

Songza

Songza plays the right music at the right time. Simply let the app know if you are at work, at the gym, relaxing, doing chores, or whatever, and it will create a playlist to fit your life at the moment. It is a pretty cool app, and using Chromecast, you can listen to the perfect music on your home's best speakers.

Songza music is curated by experts, and it is guaranteed to fit the occasion.

PostTV

Washtington. Revealed. PostTV is programming billed for Cord Cutters from the Washington Post that includes in depth videos analyzing, explaining, enlightening, and entertaining viewers. The PostTV guarantees to be smart journalism that explains, enlightens and entertains.

You can choose from Top News, Sports, Politics, Features, and even Live video from the Washington Post. The PostTV political shows include "In Play" and "On Background," and its nightly news video is called "The Fold." Access these and so much more using PostTV from your Chromecast.

Viki

Viki is a play on the words "video" and "wiki." It features Korean dramas and Japanese anime along with movies, telenovelas, music videos, and so much more. Included on Viki are subtitles in over 160 languages, so almost anybody anywhere can access and enjoy this programming using their Chromecast.

Viki has 2 billion video streams featuring thousands of hours of video from content providers like NBC, BBC, KBS, SBS, TV Tokyo and many more.

Plex

Plex organizes all of your media no matter where you keep it. Subscribing to Plex, you can always have access to all your media on whatever TV the Chromecast is plugged into. Subscription fees for Plex start at $3.99 a month, or $29.99 a year, or $74.99 for a lifetime subscription.

Plex will also stream to every device you own including smartphones, tablets, computers, and others. You can even share media with your friends and family using amazing Plex features. Finally, using Plex Sync, you can take your media with you wherever you go without actually having to take your media with you.

RealPlayer Cloud

Use RealPlayer Cloud to watch your personal video collection on your TV using your Chromecast with no cables necessary. RealPlayer Cloud allows you to move, watch, and share your videos with friends and family no matter where you are and no matter where they are.

Share videos anywhere without downloading the RealPlayer Cloud app. No longer do you have to carry HDMI or USB cables to connect devices, and no more worrying about formatting or converting videos. Watch videos from your PC, iPad, iPhone or Android straight on your TV with Chromecast.

Google Chrome

The fourth and final "app" that is currently available with Google Chromecast is Google Chrome. Although that may have been clear in previous entries, Google makes it a point to highlight its inclusion in the offered services section through the Chromecast. This is great because you can stream your browser, and its activity, to the bigger screen for group sharing. This is an awesome way to surf the web, especially if you are searching for vacation destinations, or simply want to enjoy directions for cleaning your floors in a larger format.

Think of all the things you search through Google each day, and how many of them you would like to see in a larger format. Please do not be disgusting with your answer to that thought, but keep it light and simple so everyone thinks you are thinking about shopping for shoes on a big screen. Thank you.

In all seriousness, it is amazing to watch your fingers type on a real keyboard, while the function comes to life on the larger television screen in front of you. It feels like you have been freed from small screens and tiny data entry options for good, and you will breathe a great sigh of relief as a result.

Additional Apps

A few techie blogs and online outlets have touted Chromecast as an exceptional streaming accessory, and have mentioned that with its popularity will come a much larger app following. Google recently released its Google Cast Software Development Kit (SDK), which means that many developers will soon join the Chromecast family, and that is music to everyone's ears quite honestly. This is not surprising, as Google is definitely a brand worth hitching your wagon to, and with the limited availability they have now it was simply bound to grow into something much larger. It is yet to be determined exactly when, or what you would need to do to enjoy the upgrades, but in the spirit of all things Google, it is certain to be a user friendly upgrade.

Two new music apps that are coming for the Chromecast very soon are Rdio and Beats Music. Both companies recently announced via Twitter that they are working on developing a Chromecast app, so soon there will be even more music options for this devices.

Currently Google boasts the fact that the device updates on its own, so you are never left waiting for it to update when you simply want to watch a movie, like Apple TV often does. It will not need to reboot at any time, and should not require you to set it up more than once. Those guys (and girls) are pretty smart, and new apps automatically update without you having to purchase anything new from Google.

Tips, Tricks and Troubleshooting

The Google Chromecast works beautifully because it is simple, so there are not a ton of tips and tricks for its use yet. The best tip available is to use it with a high speed internet connection, which is usually available to just about everyone. If you have Wi-Fi, you have a broadband connection, so you should be in good shape. Keep in mind, however, that if your internet connection is so/so now, it will provide the same service when you kick the content onto the bigger screen. This means shaky video, and interrupted viewing. The Chromecast can only do as much as your connection will allow, so enjoy it that way.

The good news is, it can do it anywhere. This is great news for traveling families who have a Netflix account, as they can literally stream kid's movies, old favorites or new releases from a hotel room or rented beach house. As long as the television is compatible, and the Wi-Fi exists, you are in business.

Dragging Files from Desktop into Chrome

Google is bringing their internet browser Chrome to life – even more so than it already was – by providing the ability to drag and drop files directly into the browser. First things first, however, you must be operating Google Chrome 21 (which is in the developmental stage). This is an HTML 5 feature, and requires coding capabilities that are too advanced for the general use of the device and purposes of this guide. Suffice it to say, Google is going to perfect the art of file dragging from desktop to Chrome, allowing your entire experience to be interactive soon. It's just not part of the mainstream party yet.

Chromecast for Business

Let's be clear here, so no one gets confused. There is only ONE version of Google Chromecast at the moment. This section is simply an outline of how it can be used in business settings, and not an explanation of how the Chromecast Business version operates. Now that everyone is on the same page, this is what Google Chromecast brings to the business world: Optimal Transportability.

First, the device is an incredibly compact piece of technology that allows for easy travel, so you are not lugging a projector around from presentation to presentation. Since you can use your tablet or smartphone to transmit the information onto a larger screen, you can also leave your laptop at the office and never worry about the proper connections being available at the convention center, or it freezing altogether during your pitch. There is no better way to lose a crowd than for your well-rehearsed presentation to freeze mid-way through.

Google Chromecast allows you to share all of your documents that are saved on a Google Drive directly to the larger screen. Google Drive is a free service that the tech giant offers, and it can even access shared files from the office with ease. That is great news if you are running behind, and simply need to have someone update a document for you so you can share it at your meeting.

This technology is not simply germane to Google Docs or Google Drive, wither. In fact, you can also display native Microsoft documents by converting them to HTML, or by simply opening them in the Office Web Apps in Chrome. Imagine taking all of your word documents, PDF files and Excel Spreadsheets and projecting them to a crowd with a device that fits into your pocket.

You can also display all of your Chrome tabs to the larger screen, making it a perfect presentation resource if you wanted to share competitor web pages with the others in a conference room, or pitch a possible website of your own to the masses. The fact that it is only $35, compact and easy to carry makes it an essential tool for any business person, especially those who travel and could use the projection capability on the fly.

Reducing Playback Quality to Conserve Bandwidth

It is possible to reduce your playback quality in an effort to conserve bandwidth. This is important for those who want to effortlessly get through programming, without stuttering media or buffering issues.

In the upper right-hand corner, hit the options button to reveal Google Cast Extension Options.

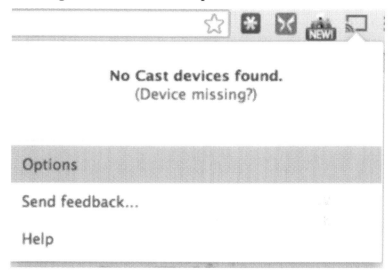

You will see the option for "Tab Projection Quality" where three viewing choices are listed:

- Extreme: 720p High Quality
- High: 720p
- Standard: 480p

High is the default setting, but the standard 480p will allow the content to stream effortlessly, while conserving bandwidth. It will provide an excellent relay of the programming, and you will not miss anything in translation. In fact, it will probably be smoother than the other two, especially if your broadband is less than efficient at certain times of the day.

Increasing Internet Speed to Increase Video Quality

While this guide is far from an official technical support for Google's device, there are some tips owners of the Chromecast might look into for smoother streaming and playback. As previously mentioned, options for the quality of the playback can be adjusted to a lower setting if the streaming seems "choppy."

Another tip is to try to increase internet speed. It's recommended that those with experience of how to configure the home internet network, or improve computer performance use these options. Remember, this guide assumes no fault for any actions that are a result of an individual adjusting things with their device, setup or computer.

Reset your modem and router – if your home network relies on just a modem, then simply power the device on and off to reset it. This simple tip may help improve internet speed alone.

If it has a router and a modem, in most cases you will unplug the router first, and then the modem. Some modems will need anywhere from 5 minutes to about a half hour before you plug them back in, or power them back on.

Try to limit devices on the network at once – The more wireless-enabled devices you have operating, the more they can drain the overall bandwidth or internet speed, sometimes. If you are trying to solely use the Chromecast, you may consider shutting other Wi-Fi devices off around the home or office.

These include laptops, smartphones, tablets, some video game consoles, printers, and other streaming media devices (such as Roku, Apple TV, etc.) If you're noticing choppy playback, this is something to test out for improving speed and making sure it is only your laptop and Chromecast using the network internet when you're streaming.

Check for spyware, viruses, etc. – Computer viruses, spyware, adware and malware an all harm internet speed or slow it down. Run any installed programs that you use to take care of these to clean up your computer. Some of the brands of "computer and internet protection" on the market include McAfee, Norton, Kaspersky and Trend Micro, all of whom offer anti-virus, spyware, firewall and other products to help out. Free options also exist online, but it is best to get the free versions from trusted websites. Running these programs regularly on the computer could help to drastically improve internet speed in many cases.

Clean up Google Chrome browser – If you use the Chrome browser to surf the internet quite a bit, then you may want to consider cleaning up the cookies, cache and browser history that get stored up. To do so, follow these instructions with an open Google Chrome browser.

1. Type "chrome://settings/clearBrowserData" into your browser's address bar without the quotes. This will take you to Chrome's "Clear browsing data" options for your internet.
2. Check off items you want to clear out. These typically include your browsing history, download history, cache, cookies and other site and plug-in data. In the most minimal cases you can clear out browsing, download history, cache and cookies.
3. Next up, on the "Obliterate the following items from:" area of the options screen, choose "the beginning of time." This will eliminate all of the aforementioned internet "stuff" or "residue" stored up by Chrome from the time you first started browsing, up until the present.
4. Once everything you want to clear is checked off, click the "Clear browsing data" button. It may take some time depending on how much "stuff" is being cleared out of your browser, so it give it some time to clean things up.
5. Once it's done, the pop-up box of options should be gone of your screen. You can close this particular tab, or load a different site on Chrome now. This may help speed up your internet and translate to better overall performance.

Check for unnecessary programs running that may use bandwidth in background – You can perform this task on either your PC or MAC computer to help free up "bandwidth hogs."

For a Windows PC, go to Settings and then "Add or Remove Programs." Uninstall any unnecessary programs you don't use or need. Keep in mind this should be done by someone with experience handling more complex computer tech issues, so that any essential items are not erased from the computer.

Pressing "Control+Alt+Delete" keys will bring up all running programs. You can use this area to shut off any programs you don't want running, that may potentially be stealing bandwidth.

On a Mac computer, you can press Command-Option-Escape keys together. This will bring up a way to force-quit applications, or you can go into the application itself to quit.

How to Get More Content Using Chromecast

Currently what you see is what you get with the Chromecast. There are four apps, including the Google's loosely referring to Chrome as an app and Netflix, Google Play and YouTube. Although there are a number of outlets like Redbox, Vimeo, and Pandora that are apparently coming aboard quickly, the best way to get more out of the Chromecast is to count on your browser for content. There is an endless amount of material on the web, so enjoy everything Chrome has to offer. There is more to come, as always, from the great minds at Google, but for now you are getting what you pay for with the $35 device. And it is well worth the expense.

Casting Browser Tabs

Since the browser option is so exciting with Chromecast, it is important to note that if any media, including music, videos, illustrations, interactive displays, and games currently play in your browser, than they will also play when cast the tabs onto your larger screen.

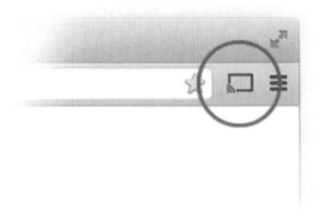

All you have to do is pull up the page or pages (tabs, as it were) and cast them onto the larger screen by clicking the "Cast" button on your Google Chrome browser. Voila! It will pop up on the larger screen and wait for your next move.

How to Cast from an App or Chrome Tab

The cast icon will show up on your YouTube, Netflix, Chrome browser (when extension is installed) or other programs or apps that interact with the Chromecast. At any time, you can click on this icon to bring up what's on your device screen on your big screen, by simply choosing to Cast it to the Chromecast. The icon will also allow you to end the casting to your big screen, and it allows for selection of certain options before you start casting.

Google Cast Extension Options

Auto-resizing – Check off this option to automatically resize your browser so it fits the TV or receiver's screen display in the best possible way when projected.

Tab projection quality – You can use this option to control the quality of the video or image output to your screen. 480 is the standard or lowest quality setting, 720p is considered "high," and 720p high bitrade is "Extreme," or the best overall quality currently supported for casting.

Fullscreen zoom – You can select a radio button for "Enabled" or "Disabled" on this option. If "Enabled" it will prevent black bars on any widescreen videos projected on the screen. If "Disabled" it will show all projected content exactly as it appears on your device screen.

Casting your Entire Computer Screen

One of the hidden "experimental" casting features allows you to project your entire computer screen display to the Chromecast so it appears on your TV or monitor. As of this publication, this feature only projects images to your TV or monitor and doesn't bring any audio with it.

To activate this screen casting feature,

1. Click on the cast icon on your Chrome Browser.
2. To the right of the bold text which says "Cast current tab" there is a small arrow for a dropdown menu. Click on that small arrow to reveal some additional options. Click "Cast entire screen (experimental)," and you will see a checkmark appear next to it.

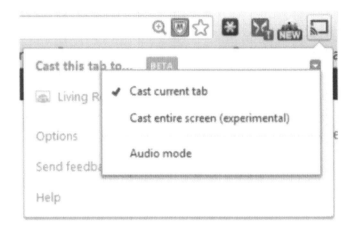

3. Select a Chromecast device to cast to.
4. Once you choose the Chromecast to case your entire screen to, you'll get a pop-up message asking if you're sure you want the app to cast your entire display. Click on "OK," and remember whatever you're showing on your computer screen will start to show up on the TV or monitor, thanks to Chromecast.

Want to stop the session? A bar will appear towards the bottom center of your screen indicating "Google Cast is sharing your screen" with a "Stop" button there. Simply click on "Stop" to stop casting your entire screen display to the Chromecast.

Content to Stream from a Chrome Tab

Some say that the Google Chrome tab casting feature may offer the most versatility with Chromecast, and that seems to be true. Being able to cast what is on an open Google Chrome browser tab brings with it the possibilities of streaming from many different websites.

Here's a list of some additional websites you might stream video content (full episodes, highlights, clips and live sports events) to your TV from:

ABC.com (Full episodes of many ABC shows)

CBS.com (Full episodes of many CBS shows)

CNET.com (Various technology-themed videos)

ESPN.com (Various ESPN video clips)

MLB.com (baseball highlights, free MLB game of day)

MTV.com (Various MTV programs including full show episodes)

MyLifeTime.com (Lifetime channel show full episodes)

NBA.com (NBA highlights & exclusives)

NBC.com (Full episodes of many NBC shows)

NBC Sports Live Extra (various live sports events: golf, tennis, etc)

Nick.com (Nickeloden kids' shows & videos)

NFL.com (NFL highlights and clips)

TVLand.com (Various full episodes of TV shows)

Watch ABC - General Hospital episodes

Vimeo.com

Most of the above websites don't require you to be a current cable or satellite TV subscriber in order to use them for online viewing. Many other sites do, such as CNBC, HGTV and DIYNetwork, so if you currently have cable or satellite, you can also enjoy content from those sites and many more!

These are just a few examples of websites with online media content that you might test out with your Chromecast to get even more enjoyment and entertainment value out of the device!

Streaming iTunes, Spotify or Other Music Programs

Spotify is a great music streaming service, which users can register and sign up for to use free. It allows for streaming all sorts of music within its vast library, so that listeners can simply search for music they want to hear and usually find it to play instantly. The Spotify program will also find any of your local music files so you can listen to them.

Chromecast doesn't allow for any sort of streaming from the Spotify app to a TV just yet. However, you can get around this by logging into your account at Spotify.com and the cast a tab from your Chrome browser to the Chromecast. You'll be able to stream your playlists and find other music to stream to your TV.

There is no built-in functionality quite yet that allows Chromecast to stream the iTunes music program to your TV display. However, there is a workaround for that, as we'll discuss below with the Chrome Remote Desktop App. It can be a bit complicated for the average user, so you may want to have someone on hand who knows there way around a computer and installing "remote access" apps.

Another option is to sign up for the Plex service which will help you stream music or other files from a Chrome tab to your Chromecast. You can find out more about this service and subscription rates at www.plexapp.com.

Audio Mode

When you are simply listening to music over through your Chromecast device, you can switch to "Audio Mode" which will adjust the bandwidth usage and frame rates, which is great if you are working while listening to music. You can make the switch effortlessly by going to your Chrome browser menu and switching into "Audio Mode."

Using Chrome Remote Desktop App

A final solution for streaming music programs to your Chromecast and TV display is to install the Chrome Remote Desktop app from the Google Chrome Store. It is free and will allow you access your computer remotely using a safe and secure log-in. As mentioned, the installation and setup of this app may be a bit complex for some users, so proceed with caution, or have someone on hand to help.

How to Setup and Use Chrome Remote Desktop

To set up the Chrome remote desktop, do the following:

1. Sign into the Gmail account you will use the most with your computer.
2. Install Chrome Remote Desktop app from Chrome store.
3. Click on green "Launch App" button, or open a new tab on your browser and go to your apps (usually by clicking on the right hand side of page until you see various apps).
4. Click on app icon on your Chrome browser to get started.
5. Follow instructions listed here at Chrome for set up and install of this to "enable remote connections."
6. You need to set up a 6-digit PIN to access your remote connection. This is essential for security of your computer and contents.
7. A message will pop up telling you it is enabling remote connections for the computer. You may get a pop up message letting you know that "Google Updates" wants to access your computer, so click "Yes" on this to continue the process.
8. You will see another pop up warning saying that Chrome Remote Desktop Access Controller wants to access your computer. Click "Yes" on this message as well.
9. You will get a pop up box with your Google email as your username along with a spot to enter the six-digit PIN you set up. Enter the PIN and click OK.
10. You should now get a message saying "Remote connections for this computer have now been enabled." It will also suggest going into your power management settings to make sure the computer won't go to sleep when it is idle. Click OK on this box and you are ready to use the Chrome remote desktop.

Next, go back to the Chrome Remote Desktop App on your browser. You should see the computer you have enabled remote access for listed there. Click on it, and enter your PIN to connect to it remotely.

Once connected, you will notice your current display window mirrored an infinite number of times cascading on your screen.

Go back up to the cast icon on your Chrome browser, and cast the tab of infinite windows to your Chromecast. It should show up on the screen.

Open a program such as Spotify, or iTunes, and you will see the program appear on your TV screen display. Now you can stream audio from your laptop computer to your TV. Keep in mind there will be a delay from the audio you hear on your laptop versus what is heard from the TV, so you may want to lower the volume completely on your computer/laptop.

At the very bottom of the screen will be a large bar indicating "Your desktop and audio output are currently shared with "[gmail user account name]" and then a "Disconnect" button. You can click "Disconnect" at any time to end your remote access session, releasing control of your computer and also ending the display on your TV monitor.

I have personally tested the Chrome Remote Desktop app using an ACER Laptop with Windows 8, and was able to project the Spotify program to stream music on my TV. I was also able to use the Chrome Remote Desktop app to stream the iTunes program and music from it on my TV, so this is decent working solution until Google brings out some new features on Chromecast.

Note: *You may notice some echo and volume fluctuation when playing music from iTunes or another program, so this may be due to streaming audio issues or network issues.*

TV as a Second Monitor?

Yes and yes. If you remember the days of Web TV, which was later purchased by Microsoft, you remember it coming with a keyboard and a remote that allowed you to browse websites and content on your television, without the actual "computer" operating system at your fingertips.

In essence, Google Chromecast provides the same luxury by turning your television into a second and larger monitor. This is perfect for working on Google documents and presentations, so you know exactly what it will look like on a larger screen, before you get in front of an audience.

It also allows you to share your pictures and videos with a living room full of people, so your Hawaii vacation videos can be revisited at once with everyone in attendance, instead of passing around your tablet while people feign interest in your tropical getaway.

This is also an awesome way to go through your kid's year in pictures with visiting parents, relatives or old friends who do not get to enjoy your social media posts as much as you would like.

Final Thoughts for the Future

At the time of this publication, Chromecast is still a very new device and technology. Since Google has released the app development software, no doubt other services will be jumping on board to work with Google to develop apps that will work with remarkable Chromecast.

In addition, there are reports from Google that iOS apps will be released to make Chromecast easier to use with iPhones and iPads. It is unclear what these apps will do, but they may unlock many new possibilities for streaming content from your device to your TV, so stay tuned!

Chromecast can also be an affordable solution for getting online entertainment in that bedroom, den, or guest room, so that you do not have to pay for a cable box for those "extra" TVs, or you do not have to buy a more expensive streaming device like Roku, Apple TV, or TiVo.

Chromecast Resources

If you have a QR code scanner app on your mobile device, you can scan our code here to go to our resource page on your mobile device. Additionally, you can go to the following URL to get helpful Chromecast resource links.

http://www.techmediasource.com/chromecast-links/

More Books by Shelby Johnson

Kindle Fire HD User's Guide Book: Unleash the Power of Your Tablet!

Facebook for Beginners: Navigating the Social Network

Kindle Paperwhite User Manual: Guide to Enjoying Your E-reader!

iPad Mini User's Guide: Simple Tips and Tricks to Unleash the Power of your Tablet!

How to Get Rid of Cable TV & Save Money: Watch Digital TV & Live Stream Online Media

Printed in Great Britain
by Amazon.co.uk, Ltd.,
Marston Gate.